RIVER TAFF
FROM SOURCE TO SEA

Alvin Nicholas &
Ingrid Nicholas

AMBERLEY

First published 2014

Amberley Publishing
The Hill, Stroud
Gloucestershire, GL5 4EP

www.amberley-books.com

Copyright © Alvin Nicholas & Ingrid Nicholas, 2014

The right of Alvin Nicholas & Ingrid Nicholas to be identified as the Author of this work has been asserted in accordance with the Copyrights, Designs and Patents Act 1988.

ISBN 978 1 4456 2087 9 (hardback)
ISBN 978 1 4456 2098 5 (ebook)

British Library Cataloguing in Publication Data.
A catalogue record for this book is available from the British Library.

Typeset in 11pt on 12pt Sabon LT Std.
Typesetting by Amberley Publishing.
Printed in the UK.

CONTENTS

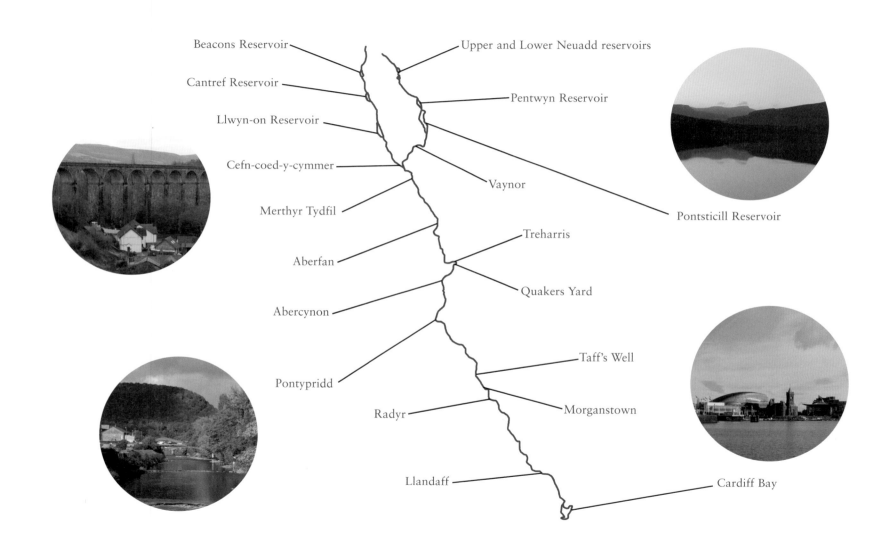

Beacons Reservoir

Upper and Lower Neuadd reservoirs

Cantref Reservoir

Pentwyn Reservoir

Llwyn-on Reservoir

Cefn-coed-y-cymmer

Vaynor

Pontsticill Reservoir

Merthyr Tydfil

Treharris

Aberfan

Quakers Yard

Abercynon

Taff's Well

Pontypridd

Morganstown

Radyr

Llandaff

Cardiff Bay

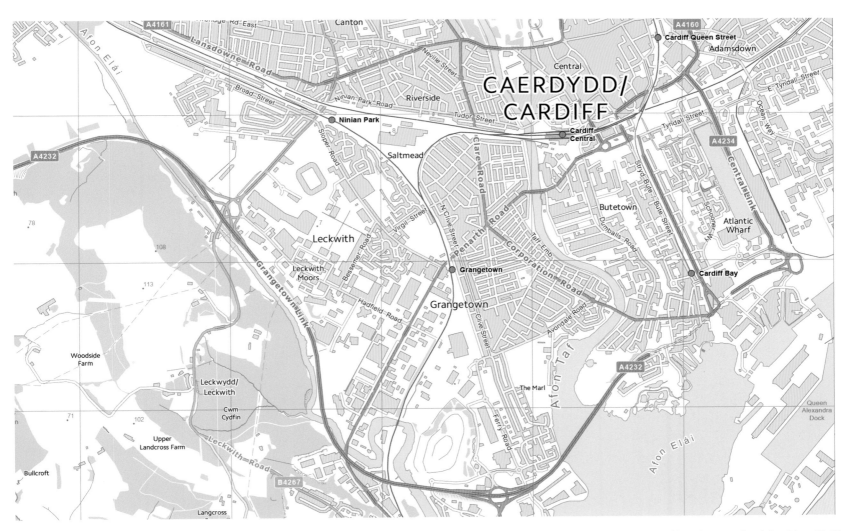

A map of Cardiff at the mouth of the River Taff

INTRODUCTION

It was a delightful walk, on a gradual slope of fifteen hundred feet in a mile and a half, with a little steeper bit at the end, and the small overhanging cap of peat at the summit. I searched over it for beetles, which were, however, very scarce, and we then walked along the ridge to the second and higher triangular summit, peeped with nervous dread on my part over the almost perpendicular precipice towards Brecon, noted the exact correspondence in slope of the two peat summits, and then back to the ridge and a little way down the southern slope to where a tiny spring trickles out – the highest source of the river Taff – and there, lying on the soft mountain turf, enjoyed our lunch and the distant view over valley and mountain to the faint haze of the Bristol Channel. We then returned to the western summit, took a final view of the grand panorama around us, and bade farewell to the beautiful mountain.

So wrote Usk-born naturalist Alfred Russell Wallace (1823–1913). Best known for independently conceiving the theory of evolution through natural selection, he was describing the genesis and receiving estuary of an iconic Welsh river. Wallace began his career as a land surveyor but went on to become an archetypal explorer-naturalist. Just two years after honing his skills in the Brecon Beacons (and having been inspired by entomologist W. H. Edward's *A Voyage up the Amazon*) he embarked upon a perilous four-year expedition to explore the Amazon and Rio Negro rivers.

Right: Overhanging cap of peat at the summit of Corn Du.

'It was a delightful walk ... with a little steeper bit
at the end...'

The Taff's twin upper reaches follow the southerly dip of the
Old Red Sandstone of the Brecon Beacons.

Keenly observant, Wallace would have noted the Taff's twin upper reaches. The Taf Fawr (Great Taff) rises as Blaen Taf Fawr on the western slopes of Corn Du, at an altitude of around 600 metres. Less than 2 kilometres to the east the Taf Fechan (Little Taff) rises as Blaen Taf Fechan below the south-east face of Pen y Fan, its highest source, as described by Wallace, appearing at an altitude of some 823 metres. The rivers Taf Fawr and Taf Fechan follow a broadly parallel course, uniting near Cyfarthfa Castle to form the River Taff (Afon Taf).

Given that a mere 60 kilometres separates source from sea, the Taff is unusually diverse. A glance at a geological map

reveals why. The twin upper reaches follow the southerly dip of the Old Red Sandstone of the Brecon Beacons. Soon, however, they encounter the northern rim of the South Wales Coalfield, where Carboniferous Limestone and Millstone Grit (officially known as the Marros Group) come to the surface. South of Merthyr the river cuts a valley through the coalfield's hilly heart before once again encountering Millstone Grit and Carboniferous Limestone – this time at the southern rim. On leaving the coalfield, the River Taff continues towards the sea amid Cardiff's relatively flat Triassic rocks.

Such geological diversity gives rise to varied and interesting landscapes. Describing a visit to the area in 1860, John Tillotson wrote,

The Taff exceeds most Welsh rivers in the wildness of its features and the vagaries of its course; it careers and leaps in rapids and cataracts, an alternate torrent and waterfall to the vicinity of Merthyr Tydfil; it proceeds with sweeping rapidity through a richly wooded glen, commanded by lofty mountains, it escapes through a channel into the luxuriant plain of Llandaff, and rolls past Cardiff to the Bristol Channel.

Despite many changes the description remains accurate. As you explore the river's course you'll find lofty peaks and wild moors, deep gorges and dramatic waterfalls. You will marvel at South Wales's coalmining and iron-making heritage and watch as the river enters the tree-cloaked Taff gorge via Wales's only thermal spring. Finally, you'll see the river meander through woods and meadows before approaching the bright lights and famous landmarks of Cardiff, terminating at the visually stunning Cardiff Bay and the Severn Estuary beyond.

The highly regarded Taff Trail follows much of the Taff's course, so exploration is both easy and enjoyable. Popular with both walkers and cyclists, it runs for 89 kilometres between Roald Dahl Plas, Cardiff Bay and Brecon, with alternative routes following the rivers Taf Fawr and Taf Fechan.

There's something new at every turn. From source to sea, the Taff will have you hooked...

Highest Source of the Taf Fechan

The tiny spring appears much as it did when Wallace visited in June 1846.

1
TWIN BEGINNINGS:
THE BRECON BEACONS TO MERTHYR TYDFIL

Established in 1957, the Brecon Beacons National Park was the tenth area in Britain to achieve National Park status. Pen y Fan (886 metres) and Corn Du (873 metres) are the highest peaks and known as the Brecon Beacons after the ancient practice of lighting signal fires. Here the mountainous topography gives rise to two upland rivers: The Taf Fawr (Great Taff) and the Taf Fechan (Little Taff). The Taf Fawr flows from a peat bog below Corn Du, while the Taf Fechan originates below the precipitous south-east face of Pen y Fan.

Both rivers have been harnessed by reservoirs, of which there are seven, completed between 1859 and 1927. The weather-blasted A470 south of Libanus utilises the Taf Fawr valley, and many will be familiar with the valley's rugged moors and beautiful, conifer-crowded reservoirs. The Taf Fechan valley is more secluded, but it's easily reached on foot from Merthyr Tydfil.

On a clear day it's possible to see the courses of both rivers from the summit of Corn Du, their respective valleys converging as they near the coalfield communities to the south.

Left: The source of the Taf Fawr.

Above and below: The Taf Fawr below Corn Du.

Magnificent cwms drop away to the north, the result of glacial action during the Pleistocene epoch, approximately 18,000 years ago. Wallace was clearly impressed:

The north-eastern summit is also triangular, a little larger than the other, and bounded by a very dangerous precipice on the side towards Brecon, where there is a nearly vertical slope of craggy rock for three or four hundred feet and a very steep rocky slope for a thousand, so that a fall is almost certainly fatal, and several such accidents have occurred, especially when parties of young men from Brecon make a holiday picnic to the summit.

The source of the Taf Fechan, below the South-East Face of Pen y Fan.

Left: Beacons Reservoir, Taf Fawr

With the opening of the West Bute Dock in 1839 and the subsequent opening of the Brunel-engineered Taff Vale Railway, Cardiff grew rapidly and the demand for water increased. The highest of the three Taf Fawr reservoirs, Beacons opened in 1897. Worked prehistoric flints have been found in the vicinity and when water levels are low, the remains of long houses can be seen in the treacherous mud.

Right: Cantref Reservoir, Taf Fawr

The second of the three Taf Fawr reservoirs was completed in 1892. Cantref today is a haven for wildlife and a great place to spot passage ospreys as they travel south during autumn. Famed for their dramatic fishing technique and boasting a 1.7-metre wingspan, these spectacular fish-eating birds of prey were persecuted in Britain from medieval times onwards, suffering particularly heavy losses at the hands of Victorian game-keepers, hunters and egg-collectors.

By 1919 they were extinct as a breeding species. A few passage migrants were seen in subsequent decades, but natural recolonisation had to wait until 1954, when a pair of Scandinavian birds famously bred at Loch Garten, 7 miles north-east of Aviemore. Since that time conservationists have made a concerted effort to nurture the species and ospreys today breed at a handful of sites in both England and Wales, in addition to their Scottish stronghold.

Llwyn-on Reservoir, Taf Fawr

The largest and southernmost of the three Taf Fawr reservoirs was completed in 1926. The old Pont-ar-Daf (Bridge over the Taff) is sometimes visible at the northern end during times of drought, a reminder of a former rural landscape submerged beneath the reservoir.

The nearby Garwnant Visitor Centre makes a good base for exploring both the reservoir and the surrounding forest, Coed Taf Fawr (Wood of the Great Taff).

The Llwyn-on dam wall.

Llwyn-on is surrounded by the beautiful Coed
Taf Fawr (Wood of the Great Taff).

Darren Fach, Site of Special Scientific Interest

Britain's rarest tree (there are around seventeen) can be found clinging to the rugged limestone cliffs of Darren Fach, just north of Merthyr. Ley's Whitebeam (or Cerdin Darren Fach in Welsh) is thought to have derived from mountain ash (*Sorbus rupicola*). It was discovered here in 1896 by amateur naturalist Revd Augustin Ley, (1842–1911), and later named after him. Darren Fach is one of only two places where this endemic species grows in the wild – the other being Penmoelallt on the opposite side of the Taf Fawr valley. A planted specimen can be seen at the Garwnant Visitor Centre.

Upper and Lower Neuadd Reservoirs, Taf Fechan

Lower Neuadd Reservoir was constructed by the Merthyr Tydfil Corporation in 1884, partly in response to leakages at Pentwyn. A nearby Victorian stone building was the filter house and is notable for its high, vaulted ceilings.

Upper Neuadd Reservoir (*above*) was completed in 1902 and at an altitude of 459 metres it is the highest in the Brecon Beacons National Park.

There are three Bronze Age cairns at the southern end of the 'island' and many flints have been found here during periods of drought including cores, scrapers, blades and arrowheads. Such finds provide evidence of our Stone Age ancestors' relationship with the landscape, and should always be reported.

Lower Neuadd Reservoir.

Many flints have been found at Upper Neuadd including cores, scrapers, blades and arrowheads.

Track leading to Lower and Upper Neuadd reservoirs, Taf Fechan valley.

Pentwyn and Pontsticill Reservoirs, Taf Fechan

Despite unprecedented industrial and urban growth (particularly between 1830 and 1860), the inhabitants of Merthyr depended upon a few wells and often resorted to travelling considerable distances to obtain spring water. Merthyr's crowded streets were, quite literally, open sewers where cholera, typhus and dysentery spread with terrifying speed.

In December 1859 the Merthyr Board of Health began constructing the first of the upper Taff reservoirs at Pentwyn (*top left*) (also known as Dolygaer Lake). Following the original dam's completion in 1863, enormous leaks of up to 11 million gallons a day occurred as it was constructed on fissured limestone and over a major geological fault known as the Neath Disturbance. Increasing industrial and domestic demand led to the construction of Lower and then Upper Neuadd reservoirs, and in 1927 a further reservoir was constructed south of Pentwyn to merge the two into the 4.2-kilometre-long Pontsticill Reservoir (*bottom left*).

The scheme required the flooding of farms, cottages, part of the old road up the valley, the original bridge and fourteenth-century Capel Taf Fechan – the remains of which can be seen during very lower water levels opposite Merthyr Tydfil Sailing Club.

The tiny Brecon Mountain Railway is based on the eastern side, and with a stunning backdrop of hills and forests, this large body of water attracts both people and wildlife. Wildfowl are present in winter and passage birds such as tufted duck, teal and mallard may be seen here.

The remains of fourteenth-century Capel Taf Fechan can be seen during very low water levels, opposite Merthyr Tydfil Sailing Club.

Opposite: In 1927 a new reservoir was constructed south of Pentwyn to merge the two into the 4.2-kilometre-long Pontsticill Reservoir.

This and next page: Bellmouth Spillway, Pontsticill Reservoir

Locally, this ominous black funnel is known as 'the plughole'. It is a superbly engineered bellmouth spillway, designed to drain excess water into a vertical shaft connected to a 235-metre-long tunnel through the base of the dam.

The Brecon Mountain Railway, Taf Fechan Valley

Construction of this narrow-gauge tourist railway began in the late 1970s utilising part of the track bed of the former Brecon and Merthyr Tydfil Junction Railway (B&MR), which opened in 1868 and was closed in 1964. Its delightful route – a great way to explore the Taf Fechan valley – currently runs for 5.6 kilometres between Pant and Pontsticill stations to Dolygaer and takes in the full length of the reservoir. Work is underway to extend the line to Torpantau, formerly the highest railway station on the B&MR.

Left: Torpantau Tunnel, Taf Fechan Valley

The B&MR was parodied as 'Breakneck and Murder' due to the high number of runaway train accidents on steep gradients such as the formidable 'Seven Mile Bank' section between Torpantau and Talybont-on-Usk (mostly 1 in 38). The 609-metre-long tunnel was the highest on the UK's standard gauge network – 400 metres above sea level at its western portal.

Next two pages: Nant y Glais, Vaynor

Small waterfalls and a group of interesting caves are associated with this Taf Fechan tributary. The stream is often intermittent due to the gorge's cavity-riddled limestone, its waters disappearing into caves and resurging lower down. Indeed, one of the finest caves to be found here is called Ogof Rhydd Sych (Cave of the Dry Ford), a reference to the oft dry stream bed. Limestone caves such as this are an important scientific and recreational resource, and much of the gorge is protected as a Site of Special Scientific Interest.

Pontsarn Bridge, Taf Fechan

Pontsarn is a contraction of Pont y Sarn Hir (Bridge of the Long Road). It's thought there was an ancient bridging point here associated with the course of a Roman road, linking the fort at Cardiff, nearby Penydarren and Y Gaer, Brecon.

Opposite: Pwll Glas, Taf Fechan

Pwll Glas (Blue Pool) is located just below the stone bridge of Pontsarn. Beloved of dreamers, artists and photographers, the pool and its attendant waterfall are utterly beguiling.

Pontsarn Station, Taf Fechan Valley

This attractive single-arched bridge carried the road to Vaynor over the railway line at Pontsarn station, of which only the platform remains. During the Victorian era the station catered for thousands of summer visitors seeking to escape from the smoke and noise of the industrial valleys, just a couple of miles down the line.

The beautiful gorge is as popular as ever. The Taff Trail conveniently follows the route of the old track, allowing visitors to explore at their leisure.

Pontsarn Viaduct, Taf Fechan

The Taf Fechan valley combines monumental manmade structures, such as Pontsarn Viaduct, with great natural beauty. Built by contractors Savin & Ward to designs by Alexander Sutherland and standing nearly 30 metres tall, this Grade II listed, seven-arched masonry viaduct carried the Brecon and Merthyr Tydfil Junction Railway across the Taf Fechan. The structure used locally quarried limestone and was opened in 1867.

Old Vaynor Church, Taf Fechan Valley

Hidden by undergrowth, a gloomy, castellated tower is all that remains of a medieval church dated to 1295. The first church to stand here was a simple wooden affair and dated back to at least the ninth century. It was burnt down during the Battle of Maesvaynor, a clash between Humphrey de Bohun, 3rd Earl of Hereford (1249–1298) and Gilbert II de Clare, Earl of Gloucester (1243–1295), after the latter erected the nearby Morlais Castle in 1287. It took the personal intervention of King Edward I to finally end the dispute.

By the latter half of the nineteenth century the thirteenth-century church had become unsafe, and in 1870 the Crawshay ironmasters commissioned the construction of a new parish church.

Vaynor Parish Church

A fourth- and final-generation Merthyr Tydfil ironmaster, Robert Thompson Crawshay (1817–1879), is buried here. His massive tombstone is inscribed with the words, 'God forgive me'.

Morlais Castle

The extensive but largely ruinous remains of Morlais Castle tower above the Taf Fechan on the vestiges of an earlier Iron Age hill fort. The castle was begun around 1287 by Gilbert II de Clare on disputed lands claimed by Humphrey de Bohun, 3rd Earl of Hereford, and is unlikely to have been completed. For such an elaborate castle there's precious little to be seen, but of particular interest is a fine, vaulted basement that has somehow survived the ravages of time.

View from Morlais Castle.

Cwm Taf Fechan Woodlands

Below Morlais Castle the river cuts through Carboniferous Limestone, eroding a spectacular narrow gorge whose sides are thick with ancient woodland. This enchanting part of the valley is both a Site of Special Scientific Interest and a Local Nature Reserve, protected for its wide variety of plants. Ash-dominated deciduous woodland on the steep slopes gives way to alder and grey willow closer to the river, along with acid grassland, limestone grassland, wet flushes and heathland. Rare species include limestone fern, green spleenwort (also a fern) and mountain melick (a lime-loving grass). In addition the humid valley is one of the best sites for mosses and liverworts in Glamorgan, with over 100 species recorded here.

Cyfarthfa Leat and Gurnos Quarry Tramroad, Taf Fechan Valley

Follow the leat (millstream) feeding Cyfarthfa Lake for just a short distance and you'll find yourself entering the beautiful Taf Fechan gorge. This well-preserved (and still functioning) artificial watercourse was built in 1825. It runs for approximately 1 kilometre from its inlet on the banks of the river to the ornamental lake at Cyfarthfa, which acted as a reservoir for the ironworks. Together, the leat and lake are reputed to have cost the Crawshays as much as the house itself.

Below the leat are the impressive remains of the Gurnos Quarry Tramroad. It was commissioned by ironmaster William Crawshay in 1792 to convey limestone using horse-drawn trams south from Gurnos Quarry to Cyfarthfa Ironworks via Pont-y-Cafnau bridge – a distance of just under 2 kilometres.

Above: Restored Section of Cyfarthfa Leat

A programme of restorative works was necessary to ensure the continued function of the Cyfarthfa leat and to prevent the adjoining walls from collapsing on to the tramroad.

Bottom left, above and next page: Paired Stone Sleepers on the Gurnos Quarry Tramroad

Look closely and you'll see the impression of the iron chairs that held the rails. The holes on either side were required to fix the chairs to the stone.

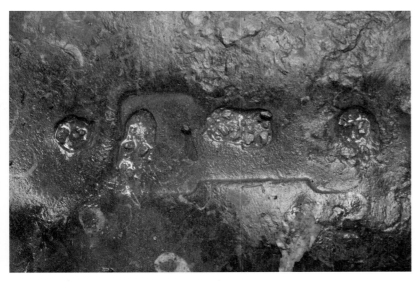

2

FIRES OF INDUSTRY:
MERTHYR TYDFIL TO TAFF'S WELL

About five miles from Merthyr, we saw in the atmosphere a faint glimmering redness appearing at intervals; as we advanced it became more fixed with occasional deeper flashes ... we could now see the men moving among the blazing fires, and hear the noise of huge hammers, clanking of chains, whiz of wheels, blast of bellows, with the deep roaring of the fires, which soon increased to a stunning degree. The effect was almost terrific when contrasted to the pitchy darkness of the night.

Account of a tour by W & S Sandys, October 1819

The fires are extinguished and the hammers silenced, but the coal and iron industry continues to influence the Taff's character between Merthyr Tydfil and Taff's Well. At every turn, one is reminded of the industrial past.

Chapel Row, Merthyr Tydfil

Merthyr Tydfil has much to offer the industrial historian. A surviving terrace of 1820s ironworkers' cottages can be seen at Chapel Row, one of which was the birthplace of musician Joseph Parry (1841–1903), best remembered for *Blodwen* (the first Welsh opera), the haunting 'Myfanwy' and many hymn tunes including 'Aberystwyth'. A partly excavated section of the Glamorganshire Canal can be seen in front of the cottages, complete with a restored and relocated iron canal bridge.

BIRTHPLACE OF
JOSEPH PARRY
1841 - 1903
MUSICIAN AND COMPOSER
'BACHGEN BACH O FERTHYR ERIOED ERIOED'

Top left: Plaque at Chapel Row.

Cyfarthfa Ironworks, Merthyr Tydfil

The first furnace was built here in 1765 by Cumbrian ironmaster Anthony Bacon (1717–1786) and his partner, William Brownrigg (1711–1800). In 1774, Crawshay dynasty founder Richard Crawshay (1739–1810) entered into the partnership, taking over Brownrigg's share. The iron-willed Yorkshireman expanded the works, eventually becoming the sole owner in 1794.

The demand for cannon during the Napoleonic Wars fuelled Cyfarthfa's early growth and its contribution was so great that in 1802 Admiral Nelson paid a personal visit, accompanied by Lord and Lady Hamilton. Moved to tears by the occasion, Crawshay roused his workers with the words, 'Here's Nelson, boys; shout you beggars!'

Cyfarthfa quickly became the largest ironworks in the world, a status it maintained until the 1830s, after which it was overtaken by its neighbour, Dowlais Ironworks. Photographs *below* and *below left* show furnace bank, Cyfarthfa Ironworks.

Rock-cut passage at the rear of the furnace bank.

Left: Cefn-coed-y-cymmer Viaduct, Taf Fawr

This magnificent fifteen-arched viaduct was built by Savin & Ward in 1866 to a sweeping curved design by engineers Alexander Sutherland and Henry Conybeare. The third largest in Wales, it carried the Brecon and Merthyr Tydfil Junction Railway over the Taf Fawr. It is 235 metres long, 36.6 metres high and today carries the Taff Trail.

Cyfarthfa Castle

Built in 1823 for William Crawshay II and set above his magnificent ironworks, Cyfarthfa Castle is Wales's best-preserved ironmaster's house. Today it houses a museum and art gallery with many artefacts relating to the industrial history of Merthyr and the colourful Crawshay dynasty.

Ornamental Pond, Cyfarthfa Castle

For the Crawshays, efficiency and profit were paramount. Even their ornamental pond had a dual purpose – it supplied water to the Cyfarthfa Ironworks.

Below left: Leat (millstream) feeding Cyfarthfa Lake.

Above: Confluence of the Taf Fechan and Taf Fawr

The rivers Taf Fechan and Taf Fawr unite at Cefn-coed-y-cymmer (Wooded Ridge at the River Confluence). 'Taf' (from the river's Welsh name, *Afon Taf)* is thought to belong to a group of similar names with a common ancient root, meaning 'to flow'. From this point on the river surges through the South Wales Coalfield and the former industrial heartlands, absorbing the rivers Taff Bargoed, Cynon and Rhondda along the way.

It seems reasonable to conclude that nineteenth-century incomers contrived 'Taffy' (the English nickname for a Welshman) to refer to Welsh people who either dwelt in or were from one of the booming Taff-side towns, and that over time the word came to be applied more generally. Not so, say experts. The prevailing view is that 'Taffy' derives from 'Daffy' (a diminutive of David), in the same way 'Paddy' is a diminutive of Patrick, and 'Jock' a diminutive of John.

Pont-y-Cafnau.

Previous page: Pont-y-Cafnau

Just below the confluence, the iron truss bridge of Pont-y-Cafnau (Bridge of Troughs) carried both the Gurnos Quarry Tramroad and an aqueduct over the River Taff to Cyfarthfa Ironworks. Dating back to 1793, it's thought to be the world's oldest cast-iron railway bridge.

Right: Richard Trevithick Monument, Merthyr Tydfil

Cornish-born inventor Richard Trevithick constructed the world's first steam locomotive to pull a load along rails at Penydarren Ironworks, Merthyr Tydfil. The locomotive ran for the first time on 13 February 1804 and on 21 February hauled 10 tonnes of iron, five wagons and seventy men 14 kilometres along the normally horse-drawn tramroad between Penydarren and Abercynon. There's a full-size replica of the famous locomotive at the National Waterfront Museum, Swansea.

Previous page: Jackson's Bridge, Merthyr

This bridge was built around 1793 to carry both the Brecon road and the Dowlais Tramroad. During the early nineteenth century it stood opposite the toughest part of town – a notorious and virtually ungovernable district known as 'China'.

This and next page: The Taff Flows through Aberfan

Like many settlements in the South Wales valleys, Aberfan grew up around its colliery and it was common practice to tip coal waste high above towns and villages rather than within the confined space of the valley bottom. At 9.15 a.m. on Friday 21 October 1966, just after registration at Pantglas Junior School, waste tip number seven of the Merthyr Vale Colliery began to move after becoming saturated by hidden springs and heavy rain. Within minutes, a farm, the school, and twenty houses had been engulfed by half a million tonnes of colliery waste. Of the 144 people who lost their lives, 116 were schoolchildren, mostly aged seven to ten.

Above: Pont y Gwaith, Edwardsville

Pont y Gwaith (Bridge of the Works) was constructed in 1811 to replace a wooden bridge associated with a nearby sixteenth-century ironworks. It shares several features in common with the famous arch bridge at Pontypridd.

Right: Merthyr Tramroad, Edwardsville

The Merthyr (or Penydarren) Tramroad ran from Merthyr to Abercynon past junctions with tramroads from Dowlais, Penydarren and Plymouth ironworks. Here at Pont y Gwaith, stone sleeper blocks, embankments and revetments remain visible.

Above: Bridge carrying the track to Pont y Gwaith over the Merthyr Tramroad.

Right: **Goitre Coed Viaduct**

Quakers Yard (so called because a group of Quakers established a burial ground here) was the location of a busy railway junction from the middle of the nineteenth century until the period after the Second World War. The Brunel-engineered Taff Vale Railway ran down the eastern bank of the Taff, crossing it via Brunel's six-arched viaduct at Goitre Coed.

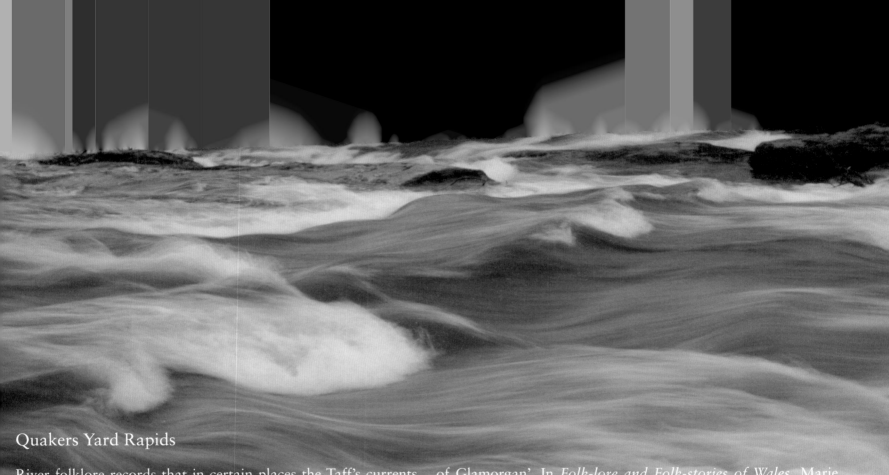

Quakers Yard Rapids

River folklore records that in certain places the Taff's currents and whirlpools harboured monsters that drowned and then devoured anyone who had the misfortune to fall in. This was one such spot. Another could once be found near Cardiff, a deep pool that was said to have been one of the 'seven wonders of Glamorgan'. In *Folk-lore and Folk-stories of Wales*, Marie Trevelyan mentions a similar pool near Merthyr. This was almost certainly Pwll Taf on the Taf Fawr, east of Cefn-coed-y-cymmer (SO 027 081). Here a resurgence from nearby caves enters the pool at a depth of 4.5 metres.

Fly Fishing on the River Taff

Improved water quality and the installation of fish passes have improved the contemplative angler's lot. Another ancient method of fishing has all but disappeared on the Taff and other Welsh rivers. Once a common sight, coracle fishing required the use of a lightweight oval boat made from an interwoven wood frame and covered in hide or fabric. A net was stretched between two coracles, which then drifted downstream taking migratory fish as they swam upstream.

Daren y Celyn Quarry

Regular travellers along the A470 will recognise this quarry, better known as the 'giant's bite' – a rocky spot that once supplied stone for local viaducts and commands wonderful views over the valleys below.

The view from Daren y Celyn Quarry.

The Cynon valley from Daren y Celyn Quarry.

The Cynon meets the Taff valley at Abercynon.

This and next page: Pontypridd

Situated at the confluence of the rivers Taff and Rhondda, the name Pontypridd is a contraction of Pont-tŷ-Pridd (Bridge of the Earthen House), since up until the late eighteenth century the area's main significance was that the river was shallow and easy to cross here. The spectacular and graceful 'New Bridge' (now known as the Old Bridge) was designed by local architect, engineer and minister William Edwards (1719–1789), and was built in 1756. It represented his fourth attempt to bridge the river. The previous three bridges were too heavy, a problem overcome by the incorporation of three cylindrical structures in each abutment. With a span of 42.6 metres, it was the longest in Britain and possibly the longest in Europe at the time. Visible in the background of the image, Victoria Bridge was built next to the Old Bridge in 1857.

Pontypridd Museum is based in the 1861 Tabernacle Chapel at the western end of Old Bridge.

Ynysangharad Park, Pontypridd

The Taff flows along the eastern side of the Edwardian-era Ynysangharad Park.

Previous page and top left: Rocking Stone, Pontypridd Common

Perched high above Ynysangharad Park there's a famous glacial boulder known as the Rocking Stone. It's surrounded by a circle of smaller standing stones constructed by nineteenth-century Arch Druid and pioneer of cremation legislation, Dr William Price (1800–1893).

Bottom left: Doctor's Bridge, Pontypridd

Doctor's Bridge (Pont y Doctor), or Machine Bridge, is a stone-built, three-arched viaduct over the River Taff, now carrying one carriageway of the A4058. It formerly carried early nineteenth-century entrepreneur Dr Richard Griffiths' Tramroad from the lower Rhondda Valley to the Glamorganshire Canal. A weighing machine for coal trams was installed at one end, hence its alternative name.

Opposite: This attractive old tramroad bridge near Taff Vale Park, Treforest, is associated with the former Taff Vale Ironworks. Today it serves as a footbridge across the River Taff.

This and next two pages: Castle Bridge and Crawshay Obelisk, Pontypridd

This three-arched, Grade II listed bridge is one of the most attractive in Pontypridd. On the west bank, south of the bridge, stands a neo-Egyptian obelisk. Oddly out of place, it is approximately 4 metres high, bears a date of 1844 and is inscribed with the initials of Francis Crawshay (1811–1878) and his brother, Henry (1812–1879), sons of William Crawshay II of Cyfarthfa Castle.

3

BREAK OUT: TAFF'S WELL TO CARDIFF

At Taff's Well the Carboniferous Limestone outcrops of Castell Coch and Lesser Garth signal a major geological change: the Taff has arrived at the coalfield's southern rim. Coal Measures give way to the younger Triassic rocks of the Glamorgan coastal plain and the river leaves the Glamorgan uplands and via the narrow Taff gorge. It finds the more congenial, undulating lowlands of the south and flows towards the capital's great parks and buildings.

Garth Hill and Lesser Garth Hill, Pentyrch

Rising to just over 1,000 feet, Garth Hill or simply 'the Garth' was the inspiration for the book and film *The Englishman Who Went Up a Hill But Came Down a Mountain* by Christopher Monger – a story about Welsh villagers building a mound on their local mountain so that English surveyors who had previously measured it and called it a hill would record it as a mountain (allegedly based upon a true incident).

The Garth is formed of resistant Pennant sandstone and from the top of the plateau there are panoramic views over the Brecon Beacons to the north, the Taff valley, Cardiff with its iconic landmarks, the Bristol Channel and distant Somerset. The prominent line of mounds are Bronze Age round barrows (graves) that have stood there for 4,000 years.

The Garth's sister hill is of equal interest. Lesser Garth is of limestone and marks the southern rim of the South Wales Coalfield. Extensively quarried, it was formerly mined for iron ore leaving many large caverns, shafts, underground lakes and tunnels.

Surrounding the enormous Taff's Well Quarry, Garth Wood Site of Special Scientific Interest (SSSI) forms a part of network of local beech woods protected as a European Special Area of Conservation (SAC). With beech growing near the western limits of its natural range, the wood exhibits a well-developed understorey and is one of the few county locations for the weird and wonderful bird's-nest orchid: a leafless, honey-coloured parasitic plant that can be found under beech trees between May and June. A population of *Porrhomma rosenhaueri* (a rare cave-dwelling spider) is present within Lesser Garth Cave, which is also notable for the discovery of artefacts dating from the Neolithic to the post-medieval period, as well as human remains spanning most of the medieval period.

The adjacent Coed y Bedw SSSI is a great place to see how geology affects vegetation. There's both Millstone Grit and Carboniferous Limestone here within a relatively small area. The Millstone Grit gives rise to acidic soils that support plants like wood sorrel, honeysuckle and native bluebell. Meanwhile, the Carboniferous Limestone gives rise to soils supporting lime-loving plants like wood false broom, dog's mercury and wild garlic. The great soil variety also supports a greater variety of trees than the surrounding woods, which are, by and large, beech dominated.

Taff's Well Quarry.

This and next page: Taff's Well Thermal Spring

It is remarkable that this natural wonder of Wales (the nation's only thermal spring) is such a little-known affair. The spring's ancient, tepid waters – once touted as a potent cure for rheumatism and other ailments – emerge on the eastern bank of the River Taff.

During the nineteenth century, Taff's Well was in great demand:

> At all hours of the day and night there are ailing and decrepit persons, men, women, and children, waiting a turn to bathe. Women must bathe here as well as men, and when a bonnet is hung on the outside, it is a sign that the gentler sex have possession.

Before the nineteenth century the spring was enclosed in iron sheets, but today it's contained within a brick-lined well and a somewhat unattractive stone building. The early history of the spring is uncertain, but it would be surprising if it were unknown to the Romans, since their settlements and roads occur throughout the Taff valley. In 1799 an 'extraordinary flood' is reported to have exposed Roman masonry that adjoined the well, but the remains were lost to the ever-encroaching river.

In common with other springs with alleged healing properties, much superstition and folklore is attached to Taff's Well. Votive offerings were placed here and the area was thought to be haunted. A ghost legend was recorded by Welsh folklorist Marie Trevelyan:

> A lady robed in grey frequently visited this well, and many people testified to having seen her in the twilight wandering along the banks of the river near the spring, or going on to the ferry under the Garth Mountain. Stories about this mysterious lady were handed down from father to son. The last was to the effect that about seventy or eighty years ago the woman in grey beckoned a man who had just been getting some of the water. He put his pitcher down and asked what he could do for her. She asked him to hold her tight by both hands until she requested him to release her. The man did as he was bidden. He began to think it a long time before she bade him cease his grip, when a 'stabbing pain' caught him in his side and with a sharp cry he loosed his hold. 'Alas! I shall remain in bondage for another hundred years, and then I must get a woman with steady hands and better than yours to hold me.' She vanished and was never seen again.

Research by the British Geological Survey suggests water enters the northern rim of the South Wales Coalfield basin (near Merthyr) and is heated by the geothermal gradient as it flows beneath the coal-bearing rocks. Under pressure due to the drop in altitude, the water eventually emerges via the Taff's Well Fault at the coalfield's southern rim some 25 kilometers to the south.

Morganstown Castle Mound or Motte

This well-preserved, low-lying motte stands on the west bank of the River Taff, directly opposite Castell Coch (itself underlain by a motte), at the southern end of the Taff gorge. Built in the late eleventh or early twelfth century prior to the Norman annexation of the Glamorgan uplands, the motte is 3.8 metres high, 30 metres in diameter at the base, has a diameter of 13 metres at the top, and is surrounded by a ditch.

These military strongholds-cum-residences were built to police newly conquered lands. This one would have been topped with a wooden tower or keep and had an adjoining bailey (a fortified enclosure) – the outline of which is visible to the east.

Castell Coch

Castell Coch (Red Castle) is quite often dismissed as a Victorian folly, but it's much more than that. Its origins can be traced to the first wave of the Norman invasion, when a large motte was built here to command the Taff gorge, perhaps in conjunction with Morganstown Motte on the other side of the river.

A stone castle was raised over the motte in the thirteenth century, and the reddish appearance of the sandstone used for its construction gave Castell Coch its name (a band of Devonian Old Red Sandstone meets Carboniferous Limestone here).

By tradition the medieval castle is associated with Ifor Bach (Little Ifor), the Welsh Lord of Senghenedd, who in 1158 overcame Cardiff Castle's defences and abducted William, Earl of Gloucester, following a dispute over land. However, Castell Coch was almost certainly raised by the Norman builder of Caerphilly Castle, Earl Gilbert II de Clare (1243–1295), soon after he annexed the upland commotes of Senghenedd in 1267.

Castell Coch was probably destroyed in the fifteenth century, and in the sixteenth century antiquary John Leyland described it thus: 'Al in Ruine no bigge thing but high.' It remained a ruin until the 1870s, when the fabulously wealthy 3rd Marquess of Bute, John Patrick Crichton-Stuart (1847–1900), commissioned the eccentric architect William Burgess to accurately restore the castle to its former glory.

Given free reign by one of the world's richest men, Burgess's imagination and taste for medieval splendour took over. The result was a truly wonderful building that combines a reconstructed medieval fortress with Victorian Gothic fantasy.

No fairytale castle would be complete without a fairytale forest. Fortunately, Castell Coch is surrounded by one. In medieval times Fforest Fawr (Great Forest) was known as Fforest Goch (Red Forest) and was probably subject to feudal forest laws.

Walnut Tree Viaduct, Taff's Well

A brick pier stands testament to a former lattice girder viaduct that spanned the gorge here. The steel structure was opened in 1901 and carried Barry Railway's link from the Rhymney Valley to Barry Docks. Following its demolition in 1969, this well-known landmark became a monument to the 1977 Silver Jubilee of Queen Elizabeth II.

Above, left and right: Ynys Bridge, Tongwynlais

Easily overlooked, this lovely nineteenth-century bridge carried the old trunk road between Morganstown and Tongwynlais. It has been superseded by a nondescript modern bridge carrying the busy B4262, a short distance upstream.

Bottom right: The Taff from the Iron Bridge

The bridge itself carried an early nineteenth-century horse-drawn tramroad (later converted to a railway) linking the Pentyrch Ironworks with the Melingriffith Tinplate Works.

The Iron Bridge.

Left: Gelynis Farmhouse

This relatively unaltered, Grade II listed, late sixteenth-century building was built for Kentish ironmaster Hugh Lambert, brought in by Sir Henry Sidney to oversee his forge at Pentyrch.

Below and next two pages: Part of the Glamorganshire Canal at Forest Farm

The Glamorganshire Canal was constructed with the backing of Merthyr's principal ironmasters and was fully completed in 1798. Prior to its construction Cardiff's coal and iron exports

were transported to the port on wagons that required four horses. A canal barge drawn by a single horse could convey twelve times the tonnage of one wagon and it immediately became economic to convey coal from distant inland collieries to the Port of Cardiff. The effect on the importance of both the Taff valley and Cardiff was dramatic, and by 1830 some 201,000 tons of coal and iron were conveyed on this canal alone.

The canal was a staggering feat of engineering. It required the construction of fifty-one locks to enable it to descend 165 metres over its 40 kilometres. There are two remaining locks on this section of the canal: Forest Lock and Middle Lock.

A mallard at Forest Farm.

This page: Melingriffith Sluice Gate and Feeder, Forest Farm

Founded in 1760, the Melingriffith Tinplate Works was one of the earliest and most important of its kind. This feeder stream provided water to the works, and the sluice gate above Radyr Weir controlled the flow of water entering the feeder.

Next page: Melingriffith Pump, Forest Farm

The recently restored Melingriffith pump was built by the eminent engineer John Rennie in 1807 to return waste water from the tinplate works to the canal.

Forest Farm and the Glamorganshire Canal Local Nature Reserve

This wonderful country park and local nature reserve is based around the eighteenth-century farmhouse and straddles the River Taff. Dippers, grey wagtail, goosander and cormorants may be seen in the river here and as water quality has improved, otter and salmon have returned to the river. At Long Wood part of the reserve is designated as a Site of Special Scientific Interest because the presence of woodland species like dog's mercury, wood anemone, native bluebell, wild garlic and yellow archangel indicate that the woodland is many centuries old.

Above and previous page: Forest Farm.

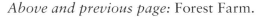

Bottom right and next page: Artificial Sand Martin Cliff and Reedbeds at Forest Farm

Sand martins nest in holes in sandy banks and feed mainly over water. An artificial cliff was constructed here because flash floods were inundating a natural sand martin nesting site on the River Taff.

Above: Fish pass, Radyr Weir.

Radyr Area

In common with many rivers in South Wales, the Taff was made virtually lifeless from industrial and other pollutants through much of the nineteenth and twentieth centuries. Today it's one of the most improved rivers in the UK, and with this has come the recovery of salmon and sea trout populations. Barriers to migration have been removed or made passable, and during October and November leaping salmon can be seen here as they travel upstream towards their spawning grounds.

The illusive Radyr hawkweed is a very rare endemic species, restricted to Radyr. It was first found in Radyr Quarry in 1907

and has been seen only a few times since. Little is known about its distribution or ecology.

Above: Reconstructed section of the Melingriffith and Pentyrch Tramway, Forest Farm.

Ty Mawr, Whitchurch

A fireplace that was removed from this house and taken to St Fagan's Castle bore a date of 1583, making this the earliest-dated surviving house in Glamorgan. It was originally a winged gentry house, but very little survives. Most of the present house was constructed in the late nineteenth century.

ON 2ND
JANUARY
1941 THE
CATHEDRAL
WAS
DEVASTATED
BY AN
ENEMY LANDMINE
WHICH FELL IN
THIS PLACE
NOW SET APART
TO RECEIVE THE
CREMATED REMAINS
OF THE
FAITHFUL DEPARTED

REMEMBER THEM
BEFORE GOD & HONOUR
THIS HALLOWED GROUND

Llandaff, Cardiff

Overlooking fertile alluvial planes and set below a steep escarpment stands historic Llandaff Cathedral. Llandaff means 'the church near the River Taff' and this hallowed spot is one of the oldest Christian sites in Britain.

A pre-Norman church that stood here was described in the twelfth-century *Book of Llandaff* and is revealed to have been very small: 8.5 metres long, 4.6 metres wide and 6.1 metres high with two aisles either side and with a porch that was 3.66 metres in length. Nothing remains of this church, but an ancient Celtic cross that stood nearby can be seen near the door of the Chapter House.

The present cathedral dates from 1107 when Urban, the first Norman Bishop, set about building a much larger church. In 1841, following centuries of damage and neglect, John Prichard (a son of the rector of Llangan) was appointed diocesan architect and oversaw the cathedral's extensive restoration. Further repairs were necessary following a Second World War bomb strike in 1941.

Right: Thirteenth-Century Stone Cross, Llandaff (Restored 1897)

Archbishop Baldwin and Giraldus Cambrensis (Gerald of Wales), preached the Third Crusade here in 1187: 'The business of the cross being publicly proclaimed at Llandaff, the English standing on one side, and the Welsh on the other, many persons of each nation took the cross.'

Below and next page: The Bishop's Palace, Llandaff

Close to the cathedral are the thirteenth-century ruins of the Bishop's Palace – a castle built for the medieval clergy of Llandaff, now a peaceful garden.

Haunted Llandaff

The tiny city of Llandaff stood outside the western boundary of Cardiff until it was made part of the capital in 1922. It has a quaint village atmosphere with many old, timbered buildings and, apparently, a well-established reputation for ghosts. In the vicinity of the river near the weir to the north of the cathedral, the sad figure of a young woman has been seen, a ghostly white lady wringing her hands in grief because her little boy drowned here. Another ghost is Bella, the wife of a local pub landlord, who many years ago committed suicide by running down the steep path next to the cathedral and jumping into the river. Her figure has been seen rushing towards the river on dark nights and disappearing into the water with a despairing cry.

Above: Llandaff Cathedral.

Right: Llandaff Fields

The heart of Cardiff boasts an unusually large area of urban green space with Llandaff Fields, Pontcanna Fields and Bute Park all situated alongside the River Taff.

Opposite: Rowers on the Taff at Cardiff.

Bute Park, Cardiff

Bute Park lies within Cardiff's historic core and boasts a wealth of attractions dating back to Roman times. The highly impressive Cardiff Castle and its famous nineteenth-century Animal Wall are but one attraction. Gorsedd stones were erected in the park in 1978 to commemorate Cardiff's hosting of the National Eisteddfod in that year and visitors can explore the site of Black Friars – a Dominican priory established during the mid-thirteenth century. The park is rich in both flora and fauna, ranging from Victorian ornamental trees to the shy and illusive otter. During autumn, the footbridge below Blackweir is another great place to spot leaping salmon.

Animal wall, Cardiff.

Bute Docks Feeder

The Bute Docks Feeder runs along the eastern edge of the park and began as a medieval mill stream. It was commissioned in the late 1830s by John Crichton-Stuart, 2nd Marquess of Bute, to replenish the first Bute (west) dock, and is still in use today.

Cardiff Bridge

Cardiff Bridge was constructed in 1859. It replaced a stone bridge of 1796 that had stood a short distance upstream but was swept away by floods in 1827.

Cardiff Castle

The first of four Roman forts was established here around AD 55 when the Romans were attempting to conquer the native Silures – who were a constant threat until their subjugation in AD 70. The final Roman fort was built of stone in around AD 280 and abandoned in the late third century AD.

A substantial motte castle was raised around 1080 within the remains of the Roman fort – and in 1126 Robert, Duke of Normandy and eldest son of William the Conqueror, died here after being imprisoned in the castle for eight years. In 1135 a twelve-sided keep was built, and further defensive improvements followed in subsequent centuries. Apartments were raised against the West Wall during the fifteenth century. This new residential core was remodelled in the Georgian and then medieval neo-Gothic styles by successive Marquesses of Bute after the castle passed into their family's hands in the mid-eighteenth century.

Millennium Stadium, Cardiff

Built for the 1999 Rugby World Cup and standing on the site of the original Cardiff Arms Park, the spectacular Millennium Stadium is the national stadium of Wales and is the home of the Wales national rugby union team. In addition it frequently hosts major football games and a wide range of other sporting, leisure and cultural events, ranging from the British Speedway Grand Prix to high-profile music concerts.

The Taff follows an artificial course past the stadium. In the early 1840s the South Wales Railway Company was searching for a site for a new railway station, but the lands here were tidal flats and the area that became Cardiff Central railway station was prone to flooding. The route of the Taff at this time was around the castle and down what is now Westgate Street.

Isambard Kingdom Brunel oversaw a major engineering scheme that saw the Taff diverted to the west and canalised, creating a larger and safer site for the station. Shipping was transferred to new docks near the mouth of the river, and the Taff was diverted away from the old quays, the location of which is commemorated by Quay Street.

Right: Railway bridge over the River Taff, Cardiff city centre with the Millennium Stadium in the background.

4

MARITIME CARDIFF:
IN AND AROUND CARDIFF BAY

The Taff, along with the River Ely, flows into Cardiff Bay to create a large artificial lake. The ambitious regeneration scheme reunited the Welsh capital with its historic waterfront, restoring the city's maritime status.

View across the Taff from Hamadryad Park

The River Taff is nearing the end of its journey.

Cardiff Bay Wetlands Reserve

This small area of freshwater marshland is located on the northern shore of Cardiff Bay near the mouth of the River Taff, adjacent to the St David's Hotel. The reserve was created in response to the disappearance of saline mudflats and the creation of a freshwater lake following the completion of the Cardiff Bay Barrage. Accessible via a gravel walkway and board walk, the open-water, reed-swamp, ditch and tall herb fen habitats support a rich assemblage of plants and animals including otter, water vole and a variety of wildfowl and other wetland birds.

Cardiff Bay

Following de-industrialisation and the development of Cardiff as an administrative centre, much of the docks became redundant. A major redevelopment of Butetown and Cardiff Docks (known locally as Tiger Bay) was launched in the late 1980s and completed in 1999. It included a barrage to enclose the estuaries of the Taff and Ely and create a 200-hectare freshwater lake. Although the barrage undoubtedly played an important role in the regeneration of the area, the controversial loss of intertidal mudflats led to a reduction in both the number and diversity of birds using the bay.

Royal Hamadryad Hospital, Cardiff Bay

HMS *Hamadryad*, a 46-gun frigate and the third ship of that name, was built at Pembroke Dockyard between 1819 and 1823, but laid up at Devonport until 1866, after which it was converted into a floating seamen's hospital for use in the Port of Cardiff. Thousands of sailors were treated aboard the ship, which was replaced by this shore-based hospital in 1897, built nearby and also called the Hamadryad.

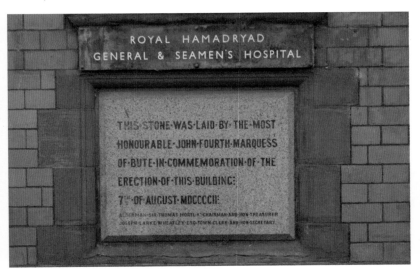

Water Sports and Boat Tours, Cardiff Bay

An extraordinary feat of engineering has turned Cardiff Bay into a haven for sailing and other water sports. The Harbour Authority is responsible for the management of the barrage and the bay, including the River Taff up to Blackweir.

A wide range of boat rides and tours allow visitors to explore more fully Cardiff Bay and the wider Severn Estuary.

The Merchant Seafarers' War Memorial, Cardiff Bay

Commissioned as a memorial to the Merchant Seamen of Cardiff Bay, Brian Fell's haunting sculpture is at once the face of every human lost forever to the sea and the forlorn hull of a ship, either beached or resting upon the ocean floor.

The Senedd, Cardiff Bay

This amazing building, designed by Richard Rogers, houses the debating chamber and committee rooms for the National Assembly for Wales.

Wales Millennium Centre, Cardiff Bay.

Left and previous page: Wales Millennium Centre, Cardiff Bay

Wales's premier arts complex is an architectural masterpiece. Two monumental inscriptions proclaim 'Creu Gwir Fel Gwydr o Ffwrnais Awen' (Creating truth like glass from inspiration's furnace) and 'In These Stones Horizons Sing'.

Opposite page: Scott Antarctic Memorial and the Norwegian Church, Cardiff Bay

Local sculptor Jonathan Williams designed this representation of Scott's epic trek towards the South Pole in 2003. The faces of his colleagues can be seen trapped in the snow and the northern end reveals the point where his expedition ship, *Terra Nova*, set sail from Cardiff in June 1910. The sculpture is pierced by a tear-shaped ice cave that Herbert Ponting photographed during the expedition.

The Norwegian church was built in 1868 to provide religious and social care to Cardiff's sizeable community of Norwegian sailors, and originally stood on the site now occupied by the Wales Millennium Centre. Born in Llandaff, Cardiff, to Norwegian parents, writer Roald Dahl was baptised in the church. Ironically, he was named after Scott's Norwegian rival, Roald Amundsen.

The Port of Cardiff

During its late nineteenth- and early twentieth-century heyday, Cardiff was the largest coal-exporting port in the world. Despite now being much smaller, the Port of Cardiff remains an important local centre for general cargo operations.

The Cardiff Bay Barrage

Separating the impounded freshwater from the saline waters of the Severn Estuary, the barrage is 800 metres long with a 300-metre-long section containing locks, sluices, bridges, and a control building. Rock stone armour protects the wave-battered seaward side, while the embankment on the bay side has been landscaped and even includes a children's park.

A fish pass that allows salmon and sea trout to return to the Taff and Ely rivers incorporates a high-tech counter that generates a silhouette of the fish, records the time and allows an estimation of length and direction of travel.

Left: Penarth Head from Cardiff Bay.

The Severn Estuary

Our journey along the River Taff ends here. Gazing seaward from the Cardiff Bay Barrage, the islands of Steep Holm and Flat Holm indicate where the Severn Estuary becomes the Bristol Channel, beyond which lies the vastness of the North Atlantic Ocean.

INDEX